THE LONG DRAG

Settle to Carlisle Portfolio

GAVIN MORRISON

LONDON

IAN ALLAN LTD

First published 1990

ISBN 0 7110 1893 6

Published by Ian Allan Ltd, Shepperton, Surrey; and printed by Ian Allan Printing Ltd at their works at Coombelands in Runnymede, England

Previous page:
'Britannia' 4-6-2 No 70010 with painted-on name, *Owen Glendower* is working hard as it approaches Kirkby Stephen station on the 1 in 100 gradient. The train is one of four returning Rugby football specials from Edinburgh to Cardiff which travelled over the line on Sunday 5 February 1967. Kingmoor motive power depot at Carlisle (12A) allocated four 'Britannias' for the occasion, but unfortunately one failed on shed, and a Class 47 was substituted. The other trains were hauled by Nos 70039 and 70003. All three locomotives were withdrawn within a few months. *Gavin Morrison*

Introduction to a Unique Railway

That great man of the Church, my friend the late Bishop Eric Treacy, once said in public that in his view there were three supreme man-made marvels in the north of England: Hadrian's Wall, York Minster and the Settle to Carlisle Railway. Treacy gained his love of railways through photography, not least on the Settle and Carlisle which he reckoned his favourite; and it somehow seems appropriate to use his well known comment by way of introducing this book, a celebration in pictures of the line, put together and largely filled by the work of another of his friends who drew his early inspiration from the late Bishop's work and even managed to get Treacy to officiate at his wedding! Thus, when I received Gavin's invitation to compose this introductory essay, I was more than delighted; it brought back many happy memories of a bit of railway whose fascination never palls, for all that has been written about it; while the fact that enthusiasts still keep coming back for more is sure indication of its truly magnetic quality.

Unlike the great railway photographers of this line, however, I came to the subject out of a sense of manifest curiosity when I was a geography student just after World War 2 and a full generation before the growth of the huge present-day interest. I, like many another Yorkshire lad, got to know the rugged splendours of the fell country through tramping over them with field notebook in hand and geological hammer in rucksack! But when my heavily-booted feet first reached Upper Ribblesdale there was amongst all the natural magnificence, this seemingly incongruous man-made element — a full-blown main line railway, clearly built to last yet apparently going nowhere in particular. Mind you, it was doing it with manifestly serious intent, or else why the huge earthworks and massively heavy trains which seemed to serve nobody on the way? I wanted to know why, my teachers could not tell me so I resolved to find out for myself.

It is not that man-made features were absent from this huge and inspiring landscape. They were, it is true, a bit dwarfed by it but the familiar dry stone walls, field barns, isolated houses and farmsteads all bore witness to a long history of human activity. However, these were the things which related to well-established local economic activity, could readily be understood and offered their own instinctive explanation, none of which quite seemed to apply to this railway. It too was quite obviously part of the landscape — and beautifully integrated with it as

Right:
Settle Junction signalbox, opened in 1913 and photographed on 10 May 1974.
D. C. Rodgers

many of the pictures in this book reveal — but it was not quite 'of' it in the immediate and intimate sense of the word. Herein lay my original curiosity and herein lies its continual fascination both to myself and others: a curiosity which I shall do my best to explain in the next few paragraphs, even though I doubt I was particularly unique in noticing this characteristic or the first to ask the simple question: why here and why so grand?

Of course, it did not take me long to appreciate that every now and then there *were* stations which attempted to serve the local communities (quite a lot of them in fact and not always too close to their respective villages!). But the fundamental mystery remained and it was not until I studied the larger story of the British railway system that the reason for its presence in these northern wildernesses began to make even remote sense at all. Thus, even though the Settle & Carlisle line can be and is taken at a variety of levels ranging from the purely emotional to the utterly practical, to really *understand* or appreciate it, one must first try to comprehend the organisation that built it: the English Midland Railway company. Without the Midland Railway there would have been no Settle & Carlisle in its present form and our present-day enjoyment of railways in the landscape would be immeasurably the poorer for it.

Conceived out of frustration in the 1860s and completed out of desperation in 1876, this railway occupied the mind of the Midland for over 10 years before it carried its first passengers; and even now one can still stand and marvel at the sheer tenacity with which the company pursued its aim. For the Midland wanted to get its trains to Scotland and was disinclined to take 'no' for an answer. Back in the late 1840s, when the British system was developing very rapidly, the fledgling Midland company (established in 1844) formed the essential central portion in the then one and only rather circuitous route from London to Scotland. In very few years time, however (1850), the Midland found itself by-passed by the more familiar East and West Coast routes and was in grave danger of degenerating into a minor provincial system unless something could be done. It had once been part of an Anglo-Scottish route and was determined to be so again.

The Midland (based on Derby) occupied some very strategic heartland but was in need of outlets for its traffic. Working arrangements, possible amalgamations and many other attempts were made over the best part of the next 20 years but they rarely if ever lasted long or proved particularly successful, and the company eventually concluded that the only solution to all its problems was to build its own lines to as many of its key outlets as possible. The year 1867 saw it reach Manchester, more or less independently and by 1868 its own route to London (St Pancras) was established, but the northern outlet remained blocked. The Midland had, in fact, managed to get as far as Ingleton from Leeds and Skipton, but here it had to hand over traffic to a rival concern, the London & North Western Railway (LNWR). The story is well recorded elsewhere but in essence, the Midland, feeling that it was still not getting a fair deal after some 10 years or more of rather acrimonious backbiting at Ingleton, eventually decided to go for its by now standard solution in this sort of situation: a new line altogether. In 1865,

Right:
The 'Peaks', later to become Classes 45 and 46, took over the Scottish expresses north of Leeds from steam in 1961, and monopolised these trains until they were withdrawn at the start of the 1976 summer timetable. Class 45 No D35 (later No 45117) in attractive green livery, passes Settle Junction on the down 'Thames-Clyde Express' on 12 March 1966. *Gavin Morrison*

Front cover:
For the fortunate photographers, who were lucky enough to be around on the Settle & Carlisle line on 4 November 1967, the day will probably be considered as one of the most memorable for visits to the line, with clear sun, early morning frost and plenty of steam-hauled freights. The crew of the Long Meg to Widnes anhydrite train, hauled by Stanier '8F' No 48090 will remember it for very different reasons, as this rundown locomotive refused to steam and the 6-mile section from Kirkby Stephen to Ais Gill Summit took around 50min. Needless to say the photographers had a field day. The picture shows the train leaving Moorcock Tunnel.
Gavin Morrison

Back cover:
Arten Gill Viaduct, 220yd long, 117ft high with 11 arches, is one of the most impressive on the line. It is necessary to be up fairly early in the morning to get pictures in this direction from the east side when the sun shines. The photographer was well rewarded for his efforts on 27 April 1982 when he captured Class 40 No 40076 heading south, with what appears to be a nuclear flask in the centre of the freight train. *John Whiteley*

goaded beyond measure, the company posted its intention of building a new line all the way to Carlisle and in 1866, got its Bill through Parliament.

The Midland was clearly serious — there had been doubts! — and this new and potentially rival proposal caused the now very concerned LNWR to offer far better terms from Ingleton. These were accepted and in 1869 the two companies promoted a second Bill, this time to abandon the whole business. Unfortunately for the Midland, it had also promised to serve local communities with its new route and there were other railways who were looking forward to using the Midland's new route for their through traffic. These various interests raised objections, the abandonment proposal was thrown out and the aggressive Midland now found itself obliged to build a line which it no longer really wanted or needed. Thus it was, more by accident than design, that a large man-made element of the landscape of the Northern Pennines was determined.

The only way the Midland could hope to compete in the circumstances was to make the alignment of the route as 'fast' as possible and it was this fact, above all, which gives the railway its unique character. Never before — and rarely since, certainly not in Britain — has a proper high speed main line been constructed through mountain country. It took the Midland 7 years to build it, a rate of progress which does not seem too bad in retrospect (especially when compared with our own sluggish rate of road building) but which was half as long again as had been expected, as indeed was the cost. But at the end of the long endeavour the line was revealed as one of the most magnificently integrated pieces of mid-Victorian engineering in these islands.

In part this was due to the nature of the Midland itself — never a company to pursue matters in half-hearted fashion once the die was cast. Everything was built on the sort of substantial scale which, while understandable enough for the major engineering structures such as the many viaducts and tunnels, was applied equally, with arguably less good reason, to the stations themselves and the many other architectural features. In fact, the company more than fulfilled its local obligations by putting no fewer than 19 stations along the route at a frequency (less than 4 miles average distance apart) which would not have disgraced a respectable outer-suburban commuter route.

It has to be said, however, that for all this grandiloquence, the Midland's thoughts were always on Carlisle and the line itself bears full witness to the fact. The crossing of a range of hills, of

Far left:
Nearly 15 miles of almost unbroken climbing at 1 in 100 faces the famous 'A3' *Flying Scotsman*, as it passes over the embankment just north of Settle Junction on 4 May 1987. The line to Carnforth can be seen dropping away in the background. As usual the A65 road seems to have gathered several cars to witness the fine sight. It is hard to believe that 23 years have passed since 'Flying Scotsman' made its first run after preservation over the Settle & Carlisle. It would be interesting to know how many journeys it has made over the line during the period.
L. A. Nixon

Centre left:
Settle station sign photographed on 21 February 1975. *D. C. Rodgers*

Left:
Situated on the platform at Settle station is this sign showing the crests of Settle and Carlisle, commemorating the opening of the line. *K. Marshall*

Below:
Humble duties for Carlisle Kingmoor-based 'Britannia' No 70029 *Shooting Star* as it crosses the A65 road at Settle on the viaduct which is variously called Marshfield, Church or Settle. The train is the 12.55 from Stourton to Carlisle, which was regularly hauled by 'Britannias' in the period 1965 to the end of 1967. The date was 18 April 1967, only 6 months before the locomotive was withdrawn, it having been allocated to Cardiff Canton when new in October 1952.
Gavin Morrison

course, was not an especially unusual problem in railway terms, but what makes the Settle & Carlisle particularly distinctive is that the gradient and alignment of the route were designed for high speed running throughout; and this was new. The ruling gradient was kept to 1:100 (maximum) and this forced the Midland to start climbing at Settle Junction and keep going until it had gained enough height to cross the watershed at the lowest reasonable altitude, a flat plateau at about 1,100ft on Ais Gill Moor at the head of both Wensleydale and the Eden valley. After this, the process was reversed in order to lose much the same height (at much the same gradient) in the descent towards Appleby and Carlisle.

The uniqueness stems from the fact that the northbound climbing is mostly complete by the time the line reaches the head of Ribblesdale and this is some 10 miles from Ais Gill. There was no point in climbing further so the Midland's engineers took their line audaciously across the heads of the Pennine valleys on a predominantly *level* alignment with huge sweeping curves. Those who saw it can never forget the sight of 90mph expresses storming along the hillside some 3-400ft above the level of the adjacent valley floors in places like Dentdale and Garsdale. It is almost equally as grand on the descent from Ais Gill to Appleby where the countryside changes but the objective remains the same. And it ensured that the stations were rarely as close to their named locations as one might have thought. Kirkby Stephen station is over 2 miles from the town and Dent more than 4 miles from the main settlement — and some 500ft higher to boot!

The Settle & Carlisle fulfilled its obligations to both the Midland and the local community until in 1923 it became part of the larger London Midland & Scottish system; but it was not until well after this latter company had, in turn, been succeeded by British Railways in 1948, that major changes began to take place to its role. Though engine classes had come and gone, carriages and wagons had undergone many changes and a few stations had closed, it was only in the 1950s and 1960s that the traditional pattern of activity began to alter in a fundamental way. Road haulage reduced local freight to a mere trickle (it had never been very high) and the few passengers became even fewer with increased car ownership. Eventually in 1970, after a 6-year rearguard action, all but two passing stations, Settle and Appleby, were closed and it was only a few years later that the through trains (the real *raison d'être* for the line, whether for passengers or freight) were either stopped altogether or reduced to a shadow of former days.

But then came a latter-day awakening — probably stimulated by the junketing in 1976 at the time of the line's centenary — and it is one of the final ironies that more recent protests about BR's expressed wish to close the line completely have mostly concentrated on the very issues which were furthest from the mind of the Midland when it built the route: the value to the local community. In the 1970s and 1980s, some reinstatement of passenger services was made and as this is being written, one has the supreme paradox that against a background of a BR wish to close a line which probably should never have been built in the first place, some stations have actually been reopened, a better service is offered than ever before and the line is attracting more passengers than for decades prior to 1970. What is more, a hard core of this new trade represents genuine local usage; something which had seemed impossible 20 years ago.

But it is as a piece of priceless heritage that the line is most often seen nowadays and it cannot be gainsaid that much of the extra revenue in recent years has come from people who are visiting the line just to see it or have come to ride on one of the many steam specials which have been a common sight since 1978. Interestingly, however, few save the more dedicated enthusiasts had even heard of the railway until more modern days and there can be few of the present generation who can actually recall from direct experience the days when all the goods yards were active and the steam-hauled local trains plied their trade for the handful of hardy souls who still remained to patronise the line; yet still they come. And one of the prime reasons must surely be the sheer engineering magnificence of the line itself.

For the more universal appreciation of this fact, we must thank above all those people who, since World War 2, have recognised the photogenic quality of this railway; which brings us to the main purpose of this book. I mentioned at the beginning the pioneering work of Eric Treacy, one of the first to see merit in 'placing' a train in the total landscape when most contemporaries were concentrating on the engine itself; and in this book, one of his modern-day successors has gone one step further by revealing it all in full colour. One might wish that colour film had been available a generation or two earlier and could have recorded events long gone by, but no matter, the landscape itself is unchanging and there is no way that this railway can be photographed without the total environment playing a key role, whether it be the rugged fell country or the more sylvan surroundings of the Eden valley.

To some folk, the presence of a railway can seem an alien feature (though this is less often stated these days given the impact of motorways) but strangely enough this never seems to have been the case with the Settle & Carlisle. More than most, it seems to live both with and within its surroundings and this could well stem from the fact that unlike most British lines, it was actually conceived and designed as a single entity and possesses remarkable visual and structural unity. It also stands as perhaps the best British tribute to the engineering skills and determination of our Victorian railway builders and this book by revealing that fact better than words can tell, should serve to indicate what some of the present-day fuss is all about! Only time will tell what the future may hold for this most celebrated railway, but if the worst ever happens, these pictures will keep it in the hearts and minds of all who knew it well.

David Jenkinson — February 1989
Knaresborough

Right:
The 16.35 Carlisle-Leeds passes Langcliffe headed by an unidentified Class 31 on 27 July 1983. Around this period the service over the line had reached its all-time low with only two trains per day in each direction, usually with a Class 31 and five coaches. *J. Livesey*

Left:
A lucky picture taken at Stainforth Sidings which connected the Craven Lime Company's works to the line and which opened in 1873. Heading south on a Long Meg to Widnes anhydrite train is '9F' 2-10-0 No 92110 as 'Britannia' Pacific No 70029 *Shooting Star* approaches on the 12.55 Stourton to Carlisle freight on 18 April 1976. *Gavin Morrison*

Above
The 11.50 Glasgow Central to Nottingham Midland headed by Class 45 No 45002 coasts down the bank, as it crosses the River Ribble at Helwith Bridge on 1 May 1982. This is the point on the 1 in 100 climb from Settle Junction to Blea Moor where the track is level for a few hundred yards.
Gavin Morrison

Left:
Inside the signalbox at Horton-in-Ribblesdale. It was opened on 9 August 1896 and closed on 1 May 1984. The picture was taken on 22 March 1974. *D. C. Rodgers*

Below:
The staff at Horton-in-Ribblesdale station in the 1950s and early 1960s kept the station in excellent condition, with immaculate flower beds, as shown in this picture taken on 16 May 1959. The station, known only as Horton until 1927, closed to passengers on 4 May 1970, with the goods facilities closing on 1 February 1965. *Gavin Morrison*

Left:
Colour slides of Midland '4F' 0-6-0s on the Settle & Carlisle are not very common. On 23 May 1959 a northbound freight produced No 44009, which was then allocated to Carlisle Kingmoor depot. It was making very slow progress, and 20min later could still be heard, having only reached somewhere around Ribblehead. Regrettably no attempt was made to obtain further pictures. *Gavin Morrison*

Left:
The danger in producing a colour album of a line such as the Settle & Carlisle, is to leave the readers with the impression that the sun shines all the time. This could not be further from the truth, as many photographers know to their cost. Saturday 25 March 1978 was a day when the weather did its worst, and preserved 'V2' No 4771 *Green Arrow* had to tackle the northbound climb in a blizzard. The photographer did well with Kodachrome 25 film to produce this result of the first steam after the lifting of the ban on steam over the Settle & Carlisle. *Hugh Ballantyne*

Right:
St Valentines Day 1970 brought its own special surprise, as that rare combination of snow and sun actually happened on the line. Class 25/3 No D7673, later No 25323 was making good progress northwards as it approached Ribblehead station. The spur into the quarry is seen in the foreground. It is only in the last few years that the daily ballast working to the quarry ceased. *Gavin Morrison*

Above:
An interesting comparison of seasons on the line. Climbing between Selside and Ribblehead, former Crosti-boilered '9F' 2-10-0 No 92021 heads north on a heavy 12.55 Stourton-Carlisle freight on 5 August 1967. These locomotives were occasionally seen on the line during the last year or two of steam working on British Railways. *Gavin Morrison*

Below:
On a glorious afternoon on 29 November 1969, Class 24 No D5097, later numbered 24097, was finding this loaded coal train rather heavy on the 1 in 100 gradient, and was travelling at less than 20mph. It was only by chance I visited the line on this particular day, and found West Coast main line diversions in operation. A 2hr period produced three Class 50s, one Class 40, two Class 47s and this Class 24. D5097 was withdrawn as early as February 1976. *Gavin Morrison*

8M18

Above:
Runpasts of the 'Cumbrian Mountain Express' steam-hauled specials have now been a feature of the tours for some time. One of the best is when the up train reverses back across Ribblehead Viaduct, and runs back into the station. '9F' 2-10-0 No 92220 *Evening Star* was photographed on 23 April 1984 propelling the stock out of the station.
Gavin Morrison

Right:
One of the most authentic double-headed combinations that the preservation movement can produce for the line is two Stanier Class 5s. On 4 April 1981, Stephenson Link motion member of the Class No 4767 *George Stephenson*, piloted No 5305 *Alderman A. E. Draper* on a northbound special across Ribblehead Viaduct. The easterly wind which was obviously blowing on this day, can make photography of steam-hauled trains very difficult on the line during the afternoon.
John Whiteley

4767
L M S

THE THAMES-CLYDE EXPRESS
46229

Left:
The 'Cumbrian Mountain Express' headboard had been changed for the 'Thames-Clyde Express' on this occasion, when Stanier Pacific No 46229 *Duchess of Hamilton* stormed north past Blea Moor on 26 March 1983.
Gavin Morrison

Above:
Saturday 2 May 1987 was another day when diversions from the West Coast main line were in operation. As is usual a light engine was sent from Carlisle to Blea Moor in case of failure, and on this occasion Class 37 No 37178 of Eastfield depot was employed. During a brief break in the clouds the sun shone on Blea Moor, although Ingleborough was still in shadow. Needless to say when Class 37 No 37135 unexpectedly arrived on the diverted Penzance-Aberdeen half an hour later, the sun had vanished. *Gavin Morrison*

Above:
The 09.07 Leeds-Carlisle leaves the 2,629yd-long Blea Moor Tunnel into sunny Dentdale with Class 45 No 45107 in charge on 28 April 1984. It is interesting to note that the train is loaded to 10 coaches, all of which are full on this occasion. *Gavin Morrison*

Right:
The south end of Dent station, 18 miles 64 chains from Settle Junction, is the setting for this picture of the up 'Cumbrian Mountain Express' as it heads south in fine style, headed by the famous Keighley & Worth Valley Railway-based 'West Country' Pacific No 34092 *City of Wells*. It was extremely fortunate that the half-minute period of sunshine just coincided with the passing of the train on 31 August 1985.
Gavin Morrison

34092

Right:
One of the most familiar photographic locations on the line was from the signalbox steps at Dent. An interesting comparison of motive power on northbound freights is shown on this and the following page, with former Crosti-boilered '9F' 2-10-0 No 92021 passing with the 12.55 Stourton to Carlisle freight on 5 August 1967. Arten Gill Viaduct can be seen further down Dentdale. *Gavin Morrison*

Above:
The sign which currently still stands above the up platform at Dent. The station opened in August 1877, is 1,145ft above sea level, and is 4 miles from the community it was built to serve. Dent village is at the bottom of the valley around 500ft below.
Gavin Morrison

Right:
The winter of 1963 was very severe in the Pennines, and the line was blocked in January and February in Dent cutting, '8F' 2-8-0 No 48149 being one of the locomotives stranded. Stanier Class 5 No 45491 passes northbound through the station on 26 January 1963, just after the line had been reopened. Remarkably, in spite of the amount of snow around, the road from the bottom of the valley to the station was just passable. *Gavin Morrison*

Right:
Class 37s have never been regular performers on the line, but on 1 April 1984 No 37096 was allocated to work this empty stock train from Carlisle to Blackpool. The location is the northerly entrance to the 1,213yd-long Rise Hill Tunnel, originally called Black Moss. The tunnel has two ventilation shafts, the deepest being 147ft, and the construction took 4 years to complete. Behind the bridge in the background was the site of the Garsdale water troughs. *K. Marshall*

Above:
A serious derailment of around 40 wagons took place in the spring of 1979 of a northbound freight at Settle Junction, resulting in a backlog of traffic. On one or two Sundays in May approximately 10 coal trains travelled the line, all Class 40 hauled, together with West Coast main line diversions. The special green-liveried Class 40 No 40106 was used on one of the northbound workings, and is passing the signalbox on 13 May 1979. *Gavin Morrison*

Right:
The snow fences above Dent station do not seem to have been very effective, judging by the amount of snow which has accumulated around the station, and which blocked the line in the cutting. Soon after reopening, Kingmoor-based Stanier Class 5 No 44669 passes northwards on a freight on 26 January 1963. *Gavin Morrison*

Above:
More West Coast main line diversions took place during the weekend of 2 May 1987. The 07.50 Manchester-Glasgow express passes Garsdale station headed by Crewe-based Class 47 No 47447. This station has over the years been called Hawes Junction, Garsdale, and Garsdale for Hawes. The station is almost 3 miles from a community of any size. The branch to Hawes opened in 1878 and closed in 1959. *John Whiteley*

Right:
A very different picture from those normally taken at Dandry Mire Viaduct showing Stanier Class 5 No 5305 *Alderman A. E. Draper* heading north on 5 March 1986 after stopping at Garsdale. *L. A. Nixon*

Above:
Dramatic lighting conditions have produced a fine picture of Class 47 No 47447 crossing Moorcock Viaduct on an up diverted West Coast main line express on 2 May 1987. It was often referred to as Dandry Mire in the early days. A viaduct was not originally considered necessary to cross the Mire, but after unsuccessful attempts to keep an embankment stable, the 12-span, 227yd-long structure was built between 1871 and 1875. *Hugh Ballantyne*

Right:
After giving the passengers another of its now legendary performances on the climb to Blea Moor, Stanier Pacific No 46229 *Duchess of Hamilton* heads its 14 coaches northwards across the 103yd-long Lunds Viaduct towards Carlisle on 29 October 1983. In the background is Moorcock Tunnel, which is 98yd long. *J. Livesey*

Far left:

A jointly run railtour from the Northeast was run in September 1963, starting on the Derwent Valley Light Railway near York, and running with Midland '4F' No 44467 to Skipton via the Wharfedale line. Stanier Pacific No 46238 *City of Carlisle*, an Upperby locomotive at the time, took over for the journey to Carlisle. The 5-coach train was hardly noticed by the Pacific which gave an uninspiring performance. The train was photographed crossing the Lunds. *Gavin Morrison*

Left:

A fine view is obtained of the line and Wild Boar Fell from above Shotlock Hill Tunnel which is only 106yd long. Class 47 No 47475 is heading south amongst the snow on 29 January 1984 on a diverted Glasgow-Paddington express.
K. Marshall

Left:
A very rare visit to the line of a Class 26 occurred on 14 April 1981 when No 26040 was used on an engineers' train. By this date the signalbox was now out of use, but the summit signboard was still in place. The down loop was still in position, but the up had been removed for some time. *L. A. Nixon*

Above right:
The fireman on 'Jubilee' No 45697 *Achilles* will be able to relax now that the summit at Ais Gill has been reached, as it is nearly all level and downhill to Leeds for the up relief to the 'Thames-Clyde Express'. Sadly the external condition of the locomotive was typical of most Kingmoor-based engines at this period, which made colour photography hardly worthwhile. The locomotive eventually became allocated to Leeds Holbeck depot, and was one of the last four 'Jubilees' to remain in service. The date of the picture is 24 July 1960. *Gavin Morrison*

Right:
The summit is in sight for the crew of Agecroft-based Fowler 'Crab' No 42726 as it heads an up freight on the evening of 13 May 1961. These locomotives were regular performers for many years over the line on freight, but were seldom seen on passenger workings. *Gavin Morrison*

Left:
One of the two summit boards, 24½ miles from Settle Junction and 48 miles from Carlisle. *Hugh Ballantyne*

Below:
The Steam Locomotive Owners' Association (SLOA) held the annual general meeting on 30 January 1983 aboard a special hauled by Gresley Pacific No 4472 *Flying Scotsman*. The locomotive was not in good form on this occasion, and made very heavy weather of the climb to Ais Gill from Appleby. The exhaust indicates a strong westerly wind as it approaches the summit. The train only went as far as Garsdale.
L. A. Nixon

Right:
The southern end of the 424yd-long Birkett Tunnel is a fine photographic location, as is shown by the picture of an unidentified Class 40 heading up the 1 in 100 gradient on 2 September 1981. In the days of steam operation, it was usually between Birkett Tunnel and Mallerstang that the gradient took its toll on a poor-steaming locomotive, and a stop for a blow-up was not unusual. *D. C. Rodgers*

1875

L N E R

Left:
A fine picture of Gresley 'A3' No 4472 *Flying Scotsman* taken on 26 July 1984 as it approaches the northern entrance to Birkett Tunnel. On this occasion it was deputising for 'A4' No 4498 *Sir Nigel Gresley*. *J. Livesey*

Above:
It took nearly 10 years to get the steam ban on the line lifted by British Railways. Like the first northbound trip, Gresley 'V2' No 4771 *Green Arrow* had the honour of working the first up special after the lifting of the ban on 27 March 1978. The train is shown climbing steadily up the 1 in 100 gradient across Birkett Common, just before the entrance to Birkett Tunnel. As can be seen, this was another case of luck being with the photographer, as the sun came out just at the right moment.
J. Livesey

Right:
Class 108 DMU leaves Kirkby Stephen with the 05.55 Skipton-Carlisle on 13 May 1987. The 1987 summer timetable showed all trains booked to stop at most stations, and there was also an early morning departure from Carlisle at 06.30. *K. Marshall*

Below right:
A very interesting scene, taken from above Kirkby Stephen station on the west side, shows an unidentified down morning express headed by Class 45 No 45113 on 28 January 1984. At this point the line is 31 miles 79 chains from Settle Junction, and the station is 2 miles from the town centre. *P. R. Walton*

Far right:
An unusual combination on the line occurred when a special from King's Cross to Howe & Co sidings was hauled by Class 25 No 25035 and Class 31 No 31102. The train is near Waitby on 5 November 1983.
K. Marshall

Right:
It is fitting that this photograph should be included in the album, as it shows the last occasion a Class 45 hauled a regular service train over the line. The location is Waitby, the locomotive No 45033, and the date 27 May 1987. It was in fact substituting for a DMU due to overcrowding, and so ended 26 years of Class 45 operation over the line.
P. R. Walton

Far right:
On 23 April 1988 Class 47 No 47583 *County of Hertfordshire* was allocated to Old Oak Common depot and found itself working the 12.37 Carlisle-Leeds. The Network SouthEast livery was well represented on the line on this day, as the following train was double-headed by a pair of Class 50s both in Network South East livery. The location is Smardale, which as the photograph shows is a superb location to observe and photograph the trains.
John Whiteley

Network SouthEast

Right:
The graceful Smardale Viaduct carries the line over the wooded valley of Scandal Beck. The 12-span structure is 237yd long, and at 130ft high was the highest on the Midland Railway. Gresley 'A4' No 60009 *Union of South Africa* coasts down the 1 in 100 gradient with the 'Cumbrian Mountain Express' on 28 April 1984. *Gavin Morrison*

Far right:
Only four coaches are needed on 29 June 1983 to carry the passengers on the 16.00 Leeds-Carlisle, headed by Class 45 No 45108, as it crosses the six-arch, 110yd-long Crosby Garrett Viaduct. Crosby Garrett station, 35¼ miles from Settle Junction, closed to passengers and goods traffic on 2 June 1952. *Gavin Morrison*

Top:
The up 'Orient Express' heads south in very dramatic lighting at Breaks Hall, headed by Class 47 No 47602 *Glorious Devon* on 15 March 1987.
P. R. Walton

Above:
Very different lighting conditions at Breaks Hall as compared to the previous view, shows a down freight headed by Class 47 No 47343 on 26 May 1982. Through freight traffic ceased on 13 May 1983. *P. R. Walton*

Right:
Another excellent location is the sweeping curve just south of the 571yd-long Helm Tunnel. A special from Leicester to Carlisle was on its return journey on 21 April 1984, with Class 45 No 45132 providing the motive power. *Gavin Morrison*

Right:
Train Tours Travel Club must have had a very interesting day out, as by the time its special had reached Appleby it had become a triple-header. The locomotives were Railfreight Class 47 No 47348, Class 45 No 45029, and most unusually Class 47/7 No 47709 *The Lord Provost*. The train was from Macclesfield to Carlisle on 28 February 1987, and any one of the locomotives would have been acceptable for a railtour by this date. Triple-headers on the line over the last few years have occurred occasionally, but not with three different liveries.
P. R. Walton

Left:
Saturday 13 May 1978 was a sad day for railway enthusiasts, as Bishop Eric Treacy, one of the finest railway photographers of all time, collapsed and died at Appleby whilst photographing '9F' 2-10-0 No 92220 *Evening Star*. As a tribute to the Bishop from enthusiasts, railwaymen and the general public, two specials were run on 30 September 1978, with a memorial service at Appleby station as shown in this picture. Three locomotives were involved; they were *Clan Line*, *Flying Scotsman* and *Evening Star*. *D. C. Rodgers*

Left:
On 28 December 1983, the last up train of the day, the 16.35 from Carlisle to Leeds pauses in the rain at Appleby hauled by an unidentified Class 47.
P. R. Walton

L M S

Left:
After the usual pause at Appleby, the down 'Cumbrian Mountain Express' leaves for Carlisle on 21 March 1987, hauled by 'Jubilee' No 5593 *Kolhapur*. This locomotive spent its last few years allocated to Holbeck depot at Leeds, and was a frequent performer over the line on relief expresses and freight trains during 1966 and 1967. *J. Livesey*

Above:
A down 'Cumbian Mountain Express' leaves Appleby at just the correct time to produce this superb silhouette of Stanier Pacific No 46229 *Duchess of Hamilton* on 29 October 1983. The photographer admits the flash of sunshine through the driving wheels was good fortune rather than skill. *K. Marshall*

Above:
This picture has been included to illustrate the change in the landscape north of Appleby compared to the south. This very rural scene shows Class 40 No D200, heading south on the 10.40 from Carlisle to Leeds on 10 August 1983. *Gavin Morrison*

Right:
1 April 1981 saw the excellent combination of the two Stanier Class 5s heading north on the 'Cumbrian Mountain Express'. In the lead is the unique member of the Class No 4767 *George Stephenson* fitted with Stephenson Link motion, which was a regular performer over the line in the last years of steam, with No 5305 *Alderman A. E. Draper* as the train engine. They are crossing the 108yd Long Marton Viaduct, which is 45 miles from Settle Junction. *L. A. Nixon*

Above:
Class 37 No 37105 was working the pick-up freight on 26 March 1986, and was photographed shunting McGhie's siding for British Gypsum at Kirkby Thore. Note the Ministry of Defence traffic on the rear of the train from Warcop. *P. R. Walton*

Right:
An unidentified Class 25 heads north past Hale Grange on a down freight. The British Gypsum factory can be seen in the background, with the McGhie's siding. This train used to follow the down passenger from Appleby at around 11.00. *K. Marshall*

Right:
Gresley 'A3' Pacific No 4472 *Flying Scotsman* bursts out of the 66yd-long Culgaith Tunnel on the up 'Cumbrian Mountain Express' on 27 July 1983. Close examination of the photograph shows that the locomotive was attached to the tender off Class A4 Pacific *Sir Nigel Gresley*.
Gavin Morrison

Far right:
Against the background of a storm, Class 47 No 47489 finds a break in the clouds as it heads south on the diverted Glasgow-Penzance express, between Langwathby and Culgaith. Both stations closed to passengers on 4 May 1970, but Langwathby reopened for Dales Rail services in 1975. *Gavin Morrison*

Right:
The '*Euston 150 Ltd*' special was run over the line on 25 July 1987. Stanier 'Princess Royal' No 6201 *Princess Elizabeth* headed the train on one of its relatively rare appearances over the line. The set of maroon coaches just gives the perfect finish to this fine picture of it heading north near Baron Wood. *J. Livesey*

Far right:
In this picture the autumn colours are at their best in the Baron Wood area, showing an 'A4' Pacific No 4498 *Sir Nigel Gresley* heading south on the up 'Cumbrian Mountain Express' on 27 October 1984. At this point the line is just over 61 miles from Settle Junction, and is now only around 330ft above sea level, having descended 870ft from Ais Gill Summit in 31 miles. *J. Livesey*

Left:
This picture has been included to show the contrast in colours around the Baron Wood area with the previous view. On 7 April 1985 the 16.35 Carlisle-Leeds was in the hands of two well turned-out Class 31/4 locomotives Nos 31464 and 31457. Such was the demand from the public by this date to ride on the line, that the trains were increased to 10 coaches, and two '31s' were needed. *Gavin Morrison*

Above:
On 26 April 1980 the 11.50 Glasgow-Nottingham had just left Armathwaite Tunnel which is 325yd long and was completed in 1871. The locomotive is Class 45 No 45024 formerly numbered D17, which was allocated to Leeds Holbeck depot when new in 1960, and was a regular performer on the line for 20 years, until withdrawn 6 months after this picture was taken in October 1980, it being one of the early withdrawals of the class. *Gavin Morrison*

Left:
Class 110 and 101 DMUs are crossing Armathwaite Viaduct working the 17.55 Carlisle-Leeds on 13 May 1989. The viaduct is 176yd long, 80ft high with nine arches. *Gavin Morrison*

Above:
An excellent restoration job was carried out on Stanier '8F' No 48151, a class which worked the line for many years, although not on passenger work, except in emergencies. Occasionally they were used to work the empty stock of the 'Starlight Expresses' north from Leeds to Carlisle, which ran in the early 1960s from Glasgow to London, if no other motive power was available. The ETHEL unit, a former Class 25 locomotive provides the train heating, but in its InterCity livery, looks rather out of place. The location is the curve south of Armathwaite station and the date 12 November 1988. *John Whiteley*

Left:
One of the 19 viaducts between Settle Junction and Carlisle, which does not appear very often in photographs is the 139yd-long Dry Beck, a nine-arch structure between Armathwaite and Low House crossing. On 27 April 1984 Class 40 No 40012 with its name *Aureol* painted on the side, heads north on the 15.55 from Leeds to Carlisle. *K. Marshall*

Above:
Low House crossing seen here, and Culgaith are the only two level crossings on the line. Class 45 No 45117 was working the 16.35 Carlisle-Leeds on 27 April 1984 past the signalbox. *K. Marshall*

Right:
Cumwhinton station closed to passengers and goods traffic on 5 November 1956, sees the 10.00 Carlisle-Leeds passing, headed by Class 45 No 45115. When this picture was taken in May 1982 the station and land were up for sale. The condition of the buildings seem excellent after such a long period since closure. The station has now been sold. *K. Marshall*

Right:
Durran Hill at Carlisle presents a very different sight since all the marshalling yards were lifted. The Midland locomotive depot was to the left of this picture showing Class 40 D200 resplendent in its green livery heading south, on what was its regular working in 1984 and 1985, the 10.00 Carlisle-Leeds, returning on the 15.55. The date of this picture is 6 March 1985.
Gavin Morrison

Above:
The splendid combination of Stanier Class 5 No 5407 and 'Jubilee' No 5690 *Leander* was photographed at Carlisle Citadel station on 29 May 1982 after working the 'Cumbrian Mountain Express' from Hellifield. It was one of those rare, glorious sunny days over the entire length of the line, which so seldom seem to coincide with the running of steam specials. In case the book leaves the reader with the impression that the sun always shines on the line, a quick word with the photographers who have kindly contributed will soon tell you otherwise. *D. C. Rodgers*

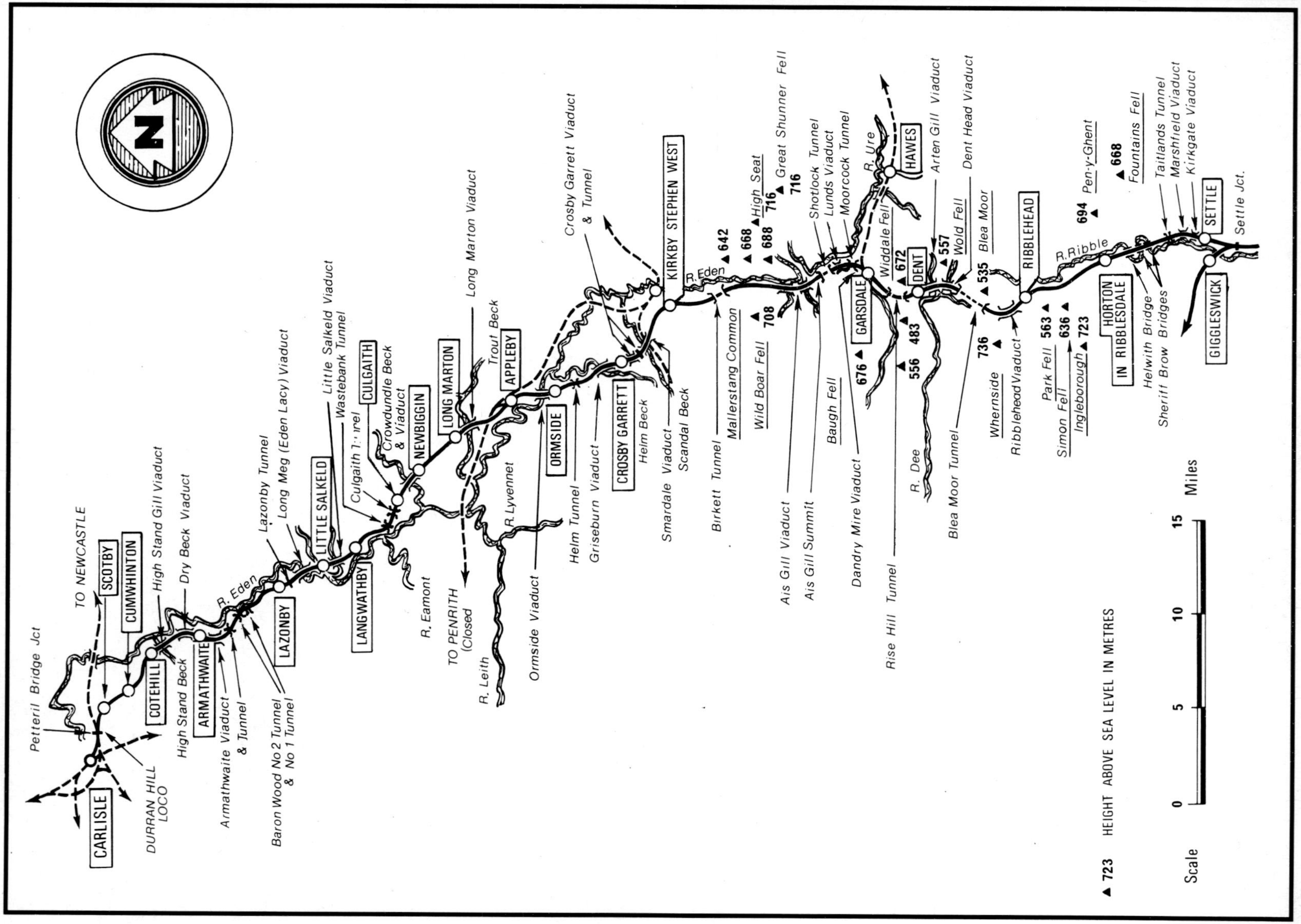

N
CARLISLE
DURRAN HILL LOCO
Petteril Bridge Jct
TO NEWCASTLE
SCOTBY
CUMWHINTON
COTEHILL
High Stand Gill Viaduct
Dry Beck Viaduct
High Stand Beck
ARMATHWAITE
Armathwaite Viaduct & Tunnel
R. Eden
Baron Wood No 2 Tunnel & No 1 Tunnel
LAZONBY
Lazonby Tunnel
Long Meg (Eden Lacy) Viaduct
LITTLE SALKELD
Little Salkeld Viaduct
Wastebank Tunnel
LANGWATHBY
Culgaith Tunnel
CULGAITH
Crowdundle Beck & Viaduct
NEWBIGGIN
R. Eamont
TO PENRITH (Closed
LONG MARTON
Long Marton Viaduct
R. Leith
Trout Beck
APPLEBY
R. Lyvennet
Ormside Viaduct
ORMSIDE
Helm Tunnel
Griseburn Viaduct
CROSBY GARRETT
Crosby Garrett Viaduct & Tunnel
Helm Beck
Smardale Viaduct
Scandal Beck
KIRKBY STEPHEN WEST
R. Eden
Birkett Tunnel
Mallerstang Common
642
668
688
708
Wild Boar Fell
High Seat
716
716
Great Shunner Fell
Ais Gill Viaduct
Ais Gill Summit
Shotlock Tunnel
Lunds Viaduct
Moorcock Tunnel
Baugh Fell
676
GARSDALE
Dandry Mire Viaduct
R. Ure
HAWES
Widdale Fell
672
DENT
483
556
Rise Hill Tunnel
R. Dee
Arten Gill Viaduct
557
Wold Fell
Dent Head Viaduct
535
Blea Moor
Blea Moor Tunnel
736
Whernside
RIBBLEHEAD
Ribblehead Viaduct
563
636
723
Park Fell
Simon Fell
Ingleborough
R. Ribble
694
Pen-y-Ghent
668
Fountains Fell
HORTON IN RIBBLESDALE
Helwith Bridge
Sheriff Brow Bridges
Taitlands Tunnel
Marshfield Viaduct
Kirkgate Viaduct
SETTLE
Settle Jct.
GIGGLESWICK
▲ 723 HEIGHT ABOVE SEA LEVEL IN METRES
Scale
0
5
10
15
Miles